THE MILLENNIUM

WEAPONS OF WAR

A Pictorial History of the Past One Thousand Years

John Hamilton

ABDO
& Daughters

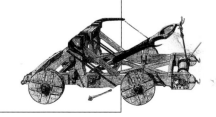

Visit us at
www.abdopub.com

Published by ABDO Publishing Company, 4940 Viking Drive, Edina, MN 55435.
Copyright ©2000 by Abdo Consulting Group, Inc. International copyrights
reserved in all countries. No part of this book may be reproduced in any form
without written permission from the publisher.

Printed in the United States.

Contributing Editors: Bob Italia, Tamara Britton, Kate Furlong
Art Direction: Pat Laurel, John Hamilton

Cover photos: AP/Wideworld Photos, Corbis
Interior photos: AP/Wideworld Photos, Corbis

Library of Congress Cataloging–in–Publication Data

Hamilton, John, 1959-
    Weapons of war / John Hamilton.
        p.cm. -- (The millennium)
    Includes index.
    Summary: In text and pictures, outlines developments in military weaponry over the
last millennium.
    ISBN 1-57765-362-9
    1. Weapons--History--Juvenile literature. [1. Weapons--History.] I. Title. II.
Millennium (Minneapolis, Minn.)

U800 .H35 2000
623.4--dc21
                                                                99-043244

CONTENTS

INTRODUCTION

More than any other human invention of the past millennium, weapons of war have helped shape civilization. Militaries with the most advanced weapons had the power to conquer lands, establish empires, and ward off enemies.

Throughout most of the millennium, close-combat weapons ruled the battlefield. But then gunpowder dramatically changed warfare, creating weapons with greater range and power.

Scientific knowledge and transportation technology greatly improved weaponry near the end of the millennium. Militaries could strike their enemies with devastating power from great distances on land, at sea, and in the air.

Land Power

Ground troops, such as infantries and cavalries, were the main military force for most of the millennium. They used daggers, swords, spears, and bows and arrows until the advent of pistols, rifles, and cannon. By the end of the millennium, powerful tanks and artillery spearheaded mobile infantries carrying automatic weapons.

Bronze- and Iron-Age metal workers forged a variety of spears and battle-axes (above). These developed into sophisticated swords and other sharp-edged weapons that were used in close combat.

Cavalries (right) were an important part of militaries until the emergence of tanks (left) in World War I.

The invention of gunpowder led to the rise of automatic rifles (above) and artillery (left).

Sea Power

Early in the millennium, wooden warships were powered by sails and oars. Steam engines, iron hulls, and powerful cannon were added as shipbuilding technology improved. By the end of the millennium, warships had grown into massive, nuclear-powered vessels capable of launching atomic weapons.

Sailing ships like the galleon (above) ruled the seas until the first ironclad ships appeared late in the millennium.

Modern supercarriers (right) are among the largest and most powerful ships in the world.

Air Power

Air power was a late invention of the millennium, beginning with the hot air balloon. But with the rapid advancement of aviation technology, air power quickly became a lethal military force. Fighters and bombers became bigger and faster, with firepower that could cripple militaries and devastate cities. By the end of the millennium, aircraft had become supersonic. Helicopters became gunships. And ballistic missiles could deliver multiple nuclear warheads to targets thousands of miles away.

The Peacekeeper ICBM (below) can deliver nuclear warheads to multiple targets.

In less than one hundred years, aircraft developed from the biplane (above, left) to the supersonic jet fighter (above, right).

BEFORE THE MILLENNIUM

The story of weapons goes back to the Stone Age, when early people used heavy rocks and clubs to defend themselves against wild beasts and rival tribes. They also crafted daggers and spearheads from flint. Stone-Age people used these kinds of weapons for hundreds of thousands of years.

Eventually, tens of thousands of years ago, early people improved their weapons by making them sail through the air. These new weapons included bows and arrows, slings, and javelins. These weapons helped ancient militaries establish huge empires.

Spears and Javelins

Spears were one of the first weapons created, and they were used throughout the world. Ancient spears were short and used for stabbing and short-distance throwing. They had sharpened, fire-hardened tips or spearheads of knapped stones. The Bronze Age saw the development of bronze spearheads.

In the Iron Age, spears became longer and had iron spearheads to combat the use of swords. Greek hoplite spears of the 6th century B.C. were up to nine feet (3 m) long. Macedonian spears were twice that length during the reign of Alexander the Great, beginning in 336 B.C. In Hellenistic times after Alexander, the spear grew to 21 feet (6 m) in length.

The Romans developed the pilum, which was the standard spear of the Roman legions.

Stone spearheads (below), often made of flint, were shaped by striking with another rock or piece of bone.

A Greek hoplite (above) with a spear

Native Americans often decorated their spears (left).

The light pilum (above, top) was a javelin, which bent or broke upon impact to prevent it from being thrown back. The heavy pilum (above, bottom) was used for thrusting and only thrown when necessary.

Bows and Arrows

By 5000 B.C., bows and arrows were being used as weapons of war. The bow and arrow, along with the spear, played an important role in battle. It allowed soldiers to attack at long range and inflict terrible losses on the enemy.

Around 1000 B.C., the composite bow was invented. It was made of wood, split horn, and dried animal tendons. This made the bow very powerful, with arrow flights up to 900 yards (823 m). Bows and arrows were a chief military weapon in Europe, the Mediterranean region, and Asia.

Before the composite bow, bows were made of wood (above).

An ancient stone painting (above) shows a battle between two rival groups of archers in Algeria.

The earliest arrowheads (above) were made of stone.

As technology progressed, metal arrowheads (above) were developed to pierce body armor.

Slings

The sling was the simplest of the ancient missile weapons. It was made of two cords or thongs fastened into a pouch. The sling's stone projectile could outdistance the javelin and some arrows. But the sling required great skill to be used effectively. It remained a weapon in ancient European militaries until the end of the Roman Empire.

Slings (right) were first used by ancient farmers and shepherds to protect their flocks from wild animals. As a weapon of war, the sling was a formidable short- and medium-range device. The slinger whirled the cords around to build up speed before letting go, releasing the projectile from its pouch.

Early Warships

The first warships were the reed boats of the ancient Egyptians. Bundles of reeds were tied together and coated with pitch to make them waterproof. Around 3000 B.C., the Egyptians were producing wooden versions of the reed boat that had oars, sails, and elevated decks for archers to shoot from.

An Egyptian reed ship

Around 2000 B.C., the Minoans of Crete recognized the difference between warships and merchant ships, and they began building ships specifically for war. Minoan ships were built long and narrow for speed, and they had a pointed bow that was used as a ram. The Greeks introduced the first seagoing warships when they began building ships covered with planking that had a frame and a keel. The ships had a bronze-covered ram, and could travel seven knots (8 mph, 13 km/h).

The Greeks continued to experiment with warship design. Their first ships were unireme ships that had one row of oarsmen. Soon, they were building biremes and triremes, ships with two and three rows of oarsmen. By the fourth century B.C., the Greeks had produced the quinquereme, with five rows of oarsmen.

Up to this point, long, skinny ships that could be rowed quickly were important, since the ram was the principal weapon for warships. In the late fourth century B.C., Demetrius I Poliorcetes of Macedonia introduced heavy missiles to ships. His ships had large catapults for hurling heavy projectiles at enemy ships.

A Greek trireme

In the first century B.C., the Romans introduced the gangplank with a hook on the end that was used to board enemy ships during battle. This allowed the Romans to use ground battle techniques at sea.

A Roman warship

Early Artillery

The ancient Greeks and Romans used a huge projectile weapon called a ballista, or siege engine, which resembled an oversized bow and arrow. Two wooden levers were drawn back under great tension and a large, arrow-like projectile was placed on a track between the two levers. When the levers were released, they sent the projectile sailing forward at high speed. Another early artillery weapon was the catapult (right), which was used to hurl large stones not only at enemy troops, but also at city walls.

A Roman catapult

Sometimes attackers placed catapults on tall siege towers (left), which allowed them to hurl stones down on the city's occupants. Archers in the siege towers also kept city defenders at bay. In the illustration at left, attackers try to breech fortified city walls with catapults, battering rams, and a siege tower, in the center right.

Greek Fire

Greek fire was an early form of chemical warfare invented in the seventh century. It was a jelly-like mixture that was highly flammable and extremely difficult to put out. It continued burning even in water, which made it a good antiship weapon.

The Greeks kept the ingredients secret for centuries, though it was probably made of sulfur and pitch in a petroleum base. Greek fire was sprayed through tubes using compressed air, or placed in pots and hurled by catapult at enemy troops or ships.

Greek fire

9

CLOSE COMBAT

Throughout the past millennium, soldiers fought using close combat. They used a variety of handheld weapons to attack their enemies. Although these weapons had a simple construction, many required much training to be used effectively.

Modern weapons gave militaries the ability to subdue the enemy at greater ranges. But at the end of the millennium, many militaries still trained their soldiers in the art of close combat.

Broadswords (left), like those favored by the European medieval knight (below), had heavy blades with very sharp edges.

Rapiers (left) were light swords with a sharp point at the tip. They were used to thrust into the enemy rather than slashing and cutting.

Guards were developed to protect the fighter's sword hand.

Swords

Swords played an important role in warfare during the Middle Ages. Medieval knights carried swords with heavy, straight, double-edged blades. Lighter swords, such as rapiers, were used for jousting. Most swords were made of iron or a combination of iron and steel.

In Europe around 1350, some swords began to have crude hand guards on the hilt. This protected the soldier's hand from his enemy's sword. By the 1500s, a wide variety of hand guards were being made.

The introduction of firearms greatly reduced the importance of swords. But many militaries continued to equip soldiers with swords in case of hand-to-hand combat. By World Wars I and II, swords had become mostly ceremonial, and were rarely used on the battlefield.

While Europeans preferred straight blades, swords in most Asian countries were curved. Samurai swords (below) are some of the most finely crafted blades in the world.

British troops (above) practice their sword-fighting skills in the mid-1800s.

A Samurai warrior

Chinese sword fighters (below) during World War II

Major General George B. McClellan (right), Commander in Chief of the Union Army 1861-1862, poses with a sword.

Spears

Advanced versions of the ancient spear were widely used in the Middle Ages.

The halberd was a six-foot (1.8 m) shaft mounted with an ax blade that had a forward point for thrusting. It was designed for fighting mounted, armored knights. But it was vulnerable against the longer lance, which a was popular cavalry weapon.

To combat the lance, the pike was developed. The pike had a hardened steel point and a long wooden shaft twice the length of the halberd. It remained a major part of European warfare until the advent of the bayonet in the late seventeenth century.

The halberd was effective against armored knights. The Japanese version of the halberd (right) is called a katana.

The lance (above) was the first strike weapon of a knight on horseback. Its power to penetrate armor was considerable. Lances ranged from 5 to 14 feet long (1.5 - 4.3 m).

Infantrymen armed with pikes (left) could easily fend off cavalry charges, even when outnumbered. Success depended upon the pikemen acting as a group.

Daggers

Daggers were shaped like miniature swords, but with a very sharp point. Most medieval knights carried them. Daggers were a weapon of last resort, but they were useful to end the pain of a fallen enemy, or to thrust through gaps in his armor.

With the advent of firearms, the importance of the dagger diminished. But it remains part of modern military weaponry in the event of hand-to-hand combat.

An ancient gold dagger found in Iran

A man wears a sheathed jambiya, *the traditional curved Islamic dagger, in his belt.*

Mace and Flails

The mace (below) was a club with a spiked or flanged metal head. The mace was widely used by warriors to smash through enemy armor.

A military flail (right) consisted of a spiked metal ball attached by a chain to a short pole. It was very effective at crushing armor.

Battle-axes

Battle-axes (left) were much like those for chopping wood. Battle-axes came in two varieties. Those with a short handle were used on horseback, and those with a long haft were used on foot. Native Americans used the tomahawk (right) as a hatchet or to throw at enemies. Asian cultures put elaborate decorations on their battle-axes.

Bows and Arrows

Bows and arrows became increasingly important in warfare in the first part of the millennium. Two types of bows, the crossbow and the longbow, were used on a large scale in warfare.

The crossbow first appeared in Europe in the 800s. The first crossbows used in warfare were made of wood and horn with bowstrings made of sinew. By the 1200s, crossbows were being made of metal. With the addition of mechanical cocking aids, crossbowmen could shoot more arrows with less effort than ever before.

During the 1100s, the longbow appeared in Europe. It was made of yew or elm and could be five to seven feet tall (1.5 - 2 m). It had bowstrings made of hemp and linen. The longbow was heavier and more difficult to use than the crossbow. But when it was used by a specially trained soldier, it could be more accurate.

Longbowmen used livery arrows for eye-level shooting and flight arrows for distance shooting. Crossbowmen used short, thick arrows called bolts. Arrowheads used in war were made of iron. They had square, blunt tips. These arrowheads were able to penetrate thick armor.

Throughout the 1400s, bows and arrows and firearms were both used in warfare. But by the early 1500s, firearms began to replace bows and arrows on the battlefield.

An archer with a longbow

Crossbow arrows, called bolts, were capable of piercing heavy armor at long distances.

A windlass handle on a small winch (below) was used to crank back the bowstring of some crossbows.

Crossbows (right) were used in China and Japan, as far back as the fourth century B.C. This powerful weapon became more common in Europe around the eleventh century. Crossbows could deliver their arrows at greater speeds and farther distances than longbows, but took longer to reload. Skilled longbowman could shoot up to 12 times per minute, but crossbows could only shoot two to three times per minute.

14

Bayonets

Maréchal de Puységur invented the bayonet in Bayonne, France, in about 1640. It replaced the pike in combat. Early bayonets had steel blades with wooden handles. The handles fit into the gun's muzzle. A gun with a bayonet could not be fired because the bayonet's handle plugged the muzzle.

A bayonet

In 1688, Sébastien Le Prestre de Vauban invented the socket bayonet. It allowed the gun to be fired with the bayonet still attached. It had a sleeve-like handle that was held in place by a stud. After Vauban's invention, bayonet design remained basically the same through the 1800s.

During World Wars I and II, bayonet blades were shortened and could be used as knives when not attached to the gun. Further improvements in repeating firearms reduced the importance of bayonets toward the end of the millennium.

Foot Chasseurs of the French Imperial Guard (above) demonstrate various tactics with the bayonet and rifle (1854).

A group of men (above) in American Revolutionary War-era military uniforms re-enact a bayonet charge.

Japanese archers (above) were specially trained and greatly skilled with their war bows, which were usually longer and more powerful than those of medieval Europe.

SHIPS

Ships enable a nation to pursue its military objectives at sea. At the beginning of the past millennium, warships were long, narrow ships that had one square sail but could also be propelled by oars. The main weaponry used on these ships was the ram and the bow and arrow. The age of sail brought more complex methods of rigging that allowed multiple sails to be used to catch wind coming from any direction. Catapults and cannon added firepower to the ships' arsenals.

The age of steam allowed ships to move without oars or sails, and the transition from wood to ironclad hulls made ships stronger. The development of even-burning smokeless powder prompted longer, bigger guns that were soon further improved by using nitroglycerin-derived explosives.

In the early twentieth century, wrought-iron hulls had been replaced by steel. World War I was the war of the battleships, with steel-hulled ships sporting twelve-inch guns (30 cm). Warships became a key factor in World War II, by deploying a new type of weapon, the airplane. At the end of the millennium, warships were powered by nuclear reactors and carried advanced weapons such as guided missiles and nuclear warheads.

The galleon was developed in the fifteenth century. It had one or two tiers of guns that were carried broadside.

Man-of-War

The seas contained the best trade routes, so the nation with the best navy gained wealth through trade. By the eighteenth century, European military ships had evolved into huge vessels supporting crews of 800 or more. A man-of-war had awesome firepower, with over 100 guns on some of the largest British and French vessels.

The British man-of-war, the HMS Victory, was launched May 7, 1765. It had 104 guns and a crew of more than 800 men.

Cannonballs were usually made of iron to destroy an enemy ship's hull, but sometimes chainshot (balls joined by a chain) was used to destroy masts. Canister shot, also called grape shot (right), was also used. It contained scores of small pistol balls designed to kill the opposing crew at close range.

Frigates, such as the 44-gun USS Constitution (left), and sloops had the advantage of speed and maneuverability over the slow-moving and clumsy men-of-war.

Ironclads

The first ironclads were ships with wooden hulls protected by iron plates. In time, ironclads were built with actual iron hulls. The first naval ship built completely of iron, the HMS *Warrior*, was launched in 1860.

On March 9, 1862, during the American Civil War, a sea battle took place that marked the beginning of a new kind of naval warfare. The Confederate *Merrimac* fought the Union *Monitor* at Hampton Roads, Virginia. This first duel between iron-protected ships ended in a draw. But the battle marked the end of wooden-hulled warships.

The Monitor *(below) and the* Merrimack *(below, right)*

Destroyers, Cruisers, and Smaller Vessels

The first U.S. Navy cruisers appeared in the late 1800s. During World War I, cruisers were used in battle. But their light armor made the ships easy to sink. During World War II, cruisers protected carriers against plane attacks. They also bombarded land before amphibious landings.

In 1893, the first destroyers appeared. During World War I, destroyers with depth charges were effective against submarines. In World War II, destroyers worked as minelayers and minesweepers. After World War II, cruisers and destroyers were fitted with missiles to protect aircraft carriers. Nuclear-powered destroyers and cruisers appeared in the early 1960s.

Motor Torpedo boats, or PT boats, were used in World War II to launch torpedoes against ships. They also carried guns.

In 1893, British shipbuilder Alfred Yarrow produced the first destroyers, the HMS Havock *(below) and the HMS* Hornet.

Torpedo boats were used extensively in World War II.

The cruiser USS Long Beach *was the world's first nuclear-powered warship.*

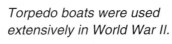

Battleships

France launched the first seaworthy battleship, *Gloire*, in 1859. It was an ironclad, steam-powered vessel with more than 30 heavy guns. Soon, other countries began to build similar battleships.

Battleship design improved through the late 1800s. Ships were outfitted with iron hulls, revolving turrets, thick armor, and more guns than ever before. Britain forever changed battleship design in 1906 with the HMS *Dreadnought*, which became the model for battleships around the world.

During World War I, battleships reached their peak. They grew larger, faster, and more powerful. Even though they did not see much combat, battleships were a key part of controlling the seas.

Battleships were less important in World War II. They were replaced by aircraft carriers because aircraft were more effective in combat than battleships.

After World War II, battleship construction stopped, and most existing battleships were decommissioned.

Dreadnought-type ships, such as the HMS Venerable, combined new technologies such as powerful steam-turbine engines, steel armor, and 12-inch guns (30 cm).

The German battleship Bismarck (left), a super-dreadnought design, was 823 feet (251 m) long. Its 15-inch (38 cm) cannon could hit targets up to 25 miles (40 km) away.

The U.S. Iowa-class ships like the USS Missouri (left) were the longest battleships ever made at 887 feet (270 m). And they were the most heavily armored, with belt armor up to 12 inches (30 cm) thick, nine 16-inch (41 cm) guns in triple turrets, and twelve 5-inch (13 cm) guns in twin turrets.

Aircraft Carriers

During World War I, the world's militaries recognized the effectiveness of the airplane as a weapon of war. On November 14, 1910, the first aircraft ever to take off from a ship left the USS *Birmingham*. On January 18, 1911, a plane landed successfully on the deck of the USS *Pennsylvania*.

Shortly after World War I, a new kind of ship was built to take advantage of naval aircraft. Aircraft carriers transported airplanes that could take off from the ship's flight deck. The planes went out to find and sink enemy ships, as well as to attack land bases. In 1918, the British Royal Navy built the first true through-deck aircraft carrier. The first true U.S. aircraft carrier was commissioned in 1934. The U.S. built fleet or attack carriers, escort carriers, and light carriers.

In 1918, the British Royal Navy converted the passenger ship HMS Argus to the first true through-deck aircraft carrier.

Aircraft carriers were key factors in World War II. The Japanese bombed Pearl Harbor from aircraft carriers. The Battle of the Coral Sea in May of 1942 was the first sea battle in which the opposing sides fought without ever coming in sight of each other. The success of the U.S. carriers against Japanese carriers in the Battle of Midway in June of 1942 was a turning point for the Allies.

After World War II, improvements to aircraft carrier construction included using nuclear energy to power most new aircraft carriers. This allowed the ships to remain at sea for longer periods of time. By the end of the millennium, aircraft carriers transported about 100 planes, as well as missiles.

The first true U.S. aircraft carrier, the USS Ranger, was commissioned in 1934.

Nimitz-class aircraft carriers are among the largest in the world. They weigh over 90,000 tons (91,440 t), and are more than 1,000 feet long (305 m). The USS Theodore Roosevelt (below) carries 90 aircraft, and has a crew of 5,617.

The first nuclear-powered aircraft carrier was the USS Enterprise (left), commissioned in 1961. The Enterprise, with eight nuclear reactors, became the largest ship in the world.

Submarines

The submarine was used in war for the first time during the American Revolution. It also made a brief appearance during the Civil War. But these early subs were small and hand-powered, and were not a factor in either war.

By World War I, submarines had become an important part of major navies worldwide. A typical submarine of World War I was powered by a diesel engine while on the surface and an electric engine while underwater. These submarines were armed with torpedoes and deck guns.

Throughout World War II, Germany led the world in submarine design. Germany's Type XXIs were outfitted with snorkels, allowing submarines to recharge their batteries without surfacing completely. Submarines played a large role in World War II until antisubmarine technology improved at the war's end.

After World War II, submarines became even more powerful with the addition of nuclear power. Nuclear power allowed submarines to stay underwater longer and travel faster than ever before.

By the end of the millennium, nuclear-powered attack submarines had become the most powerful weapons in the world. U.S. Ohio-class subs and Russian Typhoon-class subs were floating nuclear missile silos. They were difficult to detect and could stay submerged for months.

David Bushnell's Turtle *(above) was the first submarine used in war. During the American Revolution, it attacked a British ship but caused no damage. Propelled by hand-cranks, the* Turtle *could travel at 3 knots (3.4 mph, 5.5 km/h) and stay underwater for 30 minutes.*

During World Wars I and II, the German U-boat (above) was successful against supply ship convoys.

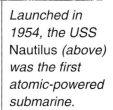

Launched in 1954, the USS Nautilus *(above) was the first atomic-powered submarine.*

U.S. Ohio-class Trident submarines (left) carry 24 missiles. Each missile carries on average five nuclear warheads for a total of about 120 warheads per submarine.

GUNS & ARTILLERY

A man plunges gunpowder into a muzzleloader of the 1770s.

The Chinese invented gunpowder in the tenth century. This invention forever changed warfare. It provided the technology for new kinds of weapons with unheard-of explosive power. These new weapons were guns and artillery.

Early guns and artillery were dangerous, inaccurate, and had to be reloaded after every shot. As technology improved, guns and artillery became easier to operate and more accurate. They continued to be an important weapon of war at the end of the millennium.

Muzzleloaders

The first guns were accident-prone and not very reliable. If there were flaws in their metal barrels, they were likely to explode in the shooter's face.

Most soldiers of the eighteenth and nineteenth centuries used muskets. These were muzzleloaders, which meant that the ammunition and gunpowder were placed down the length of the barrel and rammed down to compress them, which gave a better explosive result. These smoothbore firearms were accurate only at short range, and gave off a thick cloud of smoke when fired.

The "Brown Bess" (right), a muzzleloading flintlock musket, was the main weapon of the British Army during the American Revolution.

Matchlocks (right) were used early in the development of the gun. Gunpowder was poured into a hole in the barrel and lit with a slow-burning touch-paper. Matchlocks allowed the gunner to steady his weapon with both hands before firing.

burning wick

flash pan

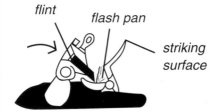

flint

flash pan

striking surface

Wheellocks and flintlocks (left) created a spark that lit the gunpowder. These types of guns were more reliable than matchlocks, though still prone to misfires.

Breechloaders

Breechloaders are firearms that are loaded from the rear. Experiments with breechloading firearms were well under way by the fifteenth century. Although they were easier to use than muzzleloaders, breechloaders often misfired.

The invention of bolt action and specially designed cartridges in the 1880s reduced the number of misfires, making the breech-loaded rifle a practical weapon. Breechloaded rifles were quick and easy to reload in battle. By the end of the millennium, breechloading was the standard way to load firearms.

B. Tyler Henry was the father of the lever action repeating rifle (right), first used in the Civil War.

The M1 Garand (left) was a gas-operated, semiautomatic rifle. It was the main rifle of the U.S. Army in World War II.

Mikhail Kalashnikov shows a model of his world-famous AK-47 assault rifle. The AK-47 is regarded as the best military rifle of the millennium. Operating as an automatic rifle, it fires at a rate of 300 to 400 rounds per minute.

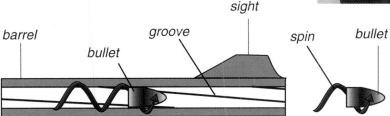

barrel · bullet · groove · sight · spin · bullet

Firearms became even more accurate after the mid-nineteenth century when the insides of barrels were rifled. Rifled barrels had spiraling grooves that make bullets spin, increasing their range and accuracy.

A close-up photo of a Smith and Wesson revolver

Revolvers

Samuel Colt patented the first successful modern revolver in 1836. Colt's revolver used a ball and powder that was loaded into the chamber from the front. A percussion cap was placed in the rear of the chamber. Pulling the trigger caused the hammer to strike the cap, which ignited the powder. The revolver was single-action. On single-action revolvers, the chamber rotates when the hammer is cocked manually.

In 1857, Horace Smith and Daniel B. Wesson produced the first revolver that used copper cartridges. In the 1870s, Smith and Wesson developed a revolver with a hinged barrel. When the gun was tipped the barrel swung away from the frame and an ejector rod expelled the spent cartridges.

Smith and Wesson's patent expired in 1872, sparking a wave of new revolver designs. The double-action revolver was developed in the 1880s. On double-action revolvers, pulling the trigger both cocks the hammer and rotates that chamber. By the end of the nineteenth century, the revolver had reached the height of its effectiveness and design. It remained much the same through the end of the millennium.

The illustration at right shows the "Peacemaker," a Colt single-action Army revolver. Revolvers were used extensively during the settling of America's western frontier.

Semiautomatic Pistols

In 1893, Ludwig Loewe & Company produced the first self-loading pistol. The Loewe pistol operated using recoil pressure. When the gun was fired, the barrel and breechblock slid back and the toggle buckled upward. This unlocked the barrel from the breechblock and allowed it to slide back. This ejected the spent case, cocked the hammer, and compressed a spring that pushed the breechblock forward. This took a fresh round from the magazine located in the grip and relocked the breechblock against the barrel.

German Georg Luger improved the spring mechanism, and American John M. Browning combined the barrel and breechblock together in a housing called a slide. The 1970s brought design innovations such as large-capacity magazines and double-action triggers.

In 1911, the U.S. Army adopted the powerful .45-caliber Colt semiautomatic pistol, designed by John M. Browning.

Georg Luger refined pistol design and produced the Luger (right), which was adopted by the German army in 1908.

Machine Guns

Machine guns are automatic weapons that use recoil or combustion gas pressure to load a round, fire it, and eject the spent cartridge. England's James Puckle patented the first machine gun in 1718. It had a revolving cylinder that fed rounds into the gun's chamber. However, the gun's flintlock ignition was not consistent enough to properly complete the load-fire-eject cycle.

The invention of the percussion cap sparked the invention of the Gatling gun in 1862 by Richard Jordan Gatling. It was used extensively during the Civil War.

The invention of smokeless powder by European chemists in the 1880s finally supplied the even combustion needed to make the machine gun's functions fully automatic. Hiram Stevens Maxim patented the first successful machine gun in 1885. In 1892, John M. Browning patented the first gas-operated machine gun.

Machine guns were used extensively during both world wars. But with the exception of air-cooled barrels and sheet-metal bodies, the machine gun operated on the same principles at the end of the millennium as it had during Hiram Maxim's time.

The Gatling gun was adopted by the U.S. military in 1865. Nicknamed "coffee grinders," Gatling guns had six or ten barrels that were cranked around by hand. It saw some service during the Civil War.

The Maxim machine gun (right) shot 600 to 700 rounds per minute. It used water as a coolant agent and had only one barrel. The recoil force of the barrel was used to perform all loading, firing, and ejecting operations. It was the principal gun in the German and Russian armies throughout World War I and accounted for most of the casualties.

The Thompson machine gun (left) was one of the first submachine guns, a shoulder weapon designed for close-range combat.

The M-60 (below) is the main machine gun of the U.S. Army.

The German MG-34 of World War II (left) could shoot 1,500 rounds per minute.

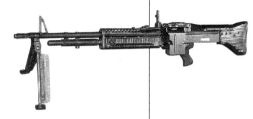

Cannon

Early cannon were developed in the 1300s. Militaries used cannon to destroy the walls of fortified cities or castles. Early cannon were muzzleloaders that had short, cone-shaped barrels made of brass or bronze. They were heavy, inaccurate, and hard to operate.

Long, tube-shaped barrels made of wrought iron appeared in the 1400s. These cannon could fire larger cannonballs with greater force and increased accuracy. When casting methods improved in the 1500s, cannon were made of cast bronze and iron. This produced lighter cannon that were easier to transport.

Basic cannon design remained the same until the late 1800s. Between 1845 and 1885, breech-loaded cannon replaced muzzleloaders, cannon barrels became rifled, and early recoil mechanisms appeared. Ammunition became more deadly with the use of explosive shells.

As its design improved, cannon became increasingly important in war. The American Civil War and World War I saw widespread use of cannon warfare. And during World War II, cannon called howitzers were used effectively on the battlefield. After World War II, many cannon were replaced by more advanced forms of artillery, such as rockets and guided missiles.

A cannon from the 1400s (above)

Early cannon (right) were loaded from the front and fired solid projectiles at targets within view.

A group of Union soldiers tends to a breechloading cannon during the Civil War (left). The most advanced cannon of the Civil War had a range of up to 2,500 yards (2.3 km).

The Paris Gun (left) was used by the Germans in World War I. It had a barrel 117 feet long (36 m), and could fire a shell at a range of up to 80 miles (129 km).

American soldiers use a camouflaged 240-mm Howitzer (right), the largest U.S. mobile field gun in use, to fire on German soldiers in western Italy during World War II.

The M109 Paladin 155mm howitzer (right) is a modern tank-like weapon that can fire many kinds of shells, including high explosive, chemical, smoke, and nuclear.

Mortars

Mortars are short, stubby cannon designed to shoot at high angles. They are important in battle because they can fire over ridges or other high obstacles. Infantry mortars are muzzleloading, indirect fire weapons with either a smooth or rifled bore and a diameter of less than 105mm. Artillery mortars have a bore of 105mm or more.

In the Middle Ages, mortars were used to shoot over castle walls, their explosive shells bursting inside the fortress. Use of mortars declined until World War I, when mortars were developed for trench warfare. A weapon was needed that could to shoot over the ridge of earth around a trench and drop down on the troops in the trench. Since 1915, small, portable models have become standard infantry weapons.

During World War II, mortars were used to keep retreated forces at bay while the offensive troops advanced into the conquered territory. Mortars were also used in the Korean and Vietnam Wars. At the end of the millennium, mortars had become essential weapons for ground troops.

Mortars were used extensively during the Civil War to bombard towns. The photo above shows Union troops manning huge mortars used during the siege of Yorktown, Virginia.

The M224 60mm mortar is used by the U.S. Army and Marine Corps. The handheld version weighs just 18 pounds (8 kg) and can fire 20 rounds per minute, at a range up to 984 yards (900 m).

A captive balloon is used by the French Republican Army for observation of enemy forces during the Battle of Fleurus, June 26, 1794.

AIRPOWER

Balloons were the first aircraft used for military purposes. They were used to observe the enemy during the French Revolutionary War and the American Civil War.

The first use of airplanes as fighters and bombers occurred during World War I. By the end of World War II, the destructive capabilities of land- and carrier-based aircraft had established airpower as an important branch of modern militaries.

Jet power, improved aerodynamic design, and exotic construction materials propelled the military airplane to new heights toward the end of the millennium. Aircraft became supersonic, capable of carrying a dizzying array of weapons and high-tech equipment very long distances.

World War I

Many of the first military aircraft were biplanes. In World War I, they were used mainly for reconnaissance. Soon, biplanes were fitted with machine guns to attack ground targets and other planes. Small bombs were also used to harass enemy soldiers. By the end of the war, more powerful engines and better construction materials made the biplane a swift and deadly fighting machine.

Airships, also known as zeppelins (above), were large, gas-filled balloons with a rigid structure underneath the skin. They were used for reconnaissance and bombing missions.

The Vickers F.B. 5 Gunbus of 1915 (left) was the first true fighter aircraft. It was a "pusher" (propeller behind the aircraft) armed with a machine gun fired by an observer who sat in front of the pilot.

The German Gotha bomber (right) was one of the first successful strategic bombers of World War I. On June 13, 1917, eighteen Gotha bombers conducted the first heavy bomber raid against London.

The Fokker D. VII's (right) high service ceiling allowed German pilots to built up speed during an attack dive. In August 1918, Fokker D. VIIs destroyed 565 Allied aircraft, making the D. VII one of the most feared aircraft of the war.

World War II

During the Second World War, faster monoplanes with retractable landing gear replaced biplanes as the chief military aircraft. Monoplanes evolved into versatile weapons of war. Fighters supported ground troops and escorted strategic bombers, which devastated many cities. Advancements continued in engine and weapon design, while all-steel body construction replaced wooden structures.

Japan's best airplane, the Mitsubishi A6M (above), was a strong and light carrier fighter with great maneuverability.

The Ju 87 "Stuka" dive-bomber (right) was crucial to Germany's early blitzkrieg tactics as it could support ground troops with pinpoint bombing.

The British Supermarine Spitfire (right) was a low-wing monoplane of all-metal, stressed-skin construction. It had eight machine guns mounted in the wings, and was Britain's best WWII fighter.

Because of its great range and speed, America's P-51 Mustang (right) was the best high-altitude escort fighter of WWII.

The Soviet Yak-9 (left) was built as a long-range fighter and ground attacker. From mid-1944, the Yak-9 was the most effective Soviet fighter.

America's B-17 Flying Fortress (right) was the war's most successful bomber. Its gun turrets for machine guns helped it battle past German fighters during long-range, daytime bombing runs.

Jet Fighters

Germany built the first operational jet-powered aircraft in 1939. Both Germany and Britain had jet fighters in service during the end of World War II, but neither saw very much action.

Turbojet propulsion replaced piston engines after the war. The first jet aircraft dogfight occurred during the Korean War. By then, most jets had swept-back wings, more powerful engines, improved aerodynamics, and high-lift devices that allowed better handling. Air-to-air missiles and radar-ranging gunsights also appeared.

The first jet-powered aircraft, the German He 178 (left), had a nose intake and a top speed of 373 mph (600 km/h). It flew for the first time on August 27, 1939.

Britain's Gloster Meteor (right) was the first jet aircraft to serve in World War II. It was effective in intercepting German V-1 rockets.

The Messerschmitt ME 262 (left) could fly 525 mph (845 km/h). It was armed with four cannon and unguided rockets. But it entered service much too late to affect the war.

The Soviet MiG-15 (left) and the North American F-86 Sabre (right) were among the first jets to engage in dogfights during the Korean War.

The Soviet MiG-21 (left) and the U.S. F-4 Phantom (right) were the premier fighters during the Vietnam War, capable of Mach 2 speeds.

Supersonic fighters, equipped with powerful afterburning engines, entered service in the mid-1950s. Improved avionics allowed most fighters to fly at night and in all weather conditions.

In the 1960s, the Mach 2 speed was broken. In the 1970s, engine acceleration improved dramatically. But performance of radars, weapons, and avionics proved even more important during combat as equipment became increasingly sophisticated.

Toward the end of the millennium, jet fighters became less specialized and more versatile, capable of land and air attacks. The biggest advance was stealth technology, a combination of exotic construction materials and radical aircraft design that allowed fighters to absorb radar energy and fly into enemy territory undetected.

The F-15 Eagle (above, left) and the F-16 Fighting Falcon (above, right) are the U.S. Air Force's main fighter aircraft. The F-15 is a single-seat, twin-engine fighter with a sophisticated radar system. The F-16 is a single-seat, multirole aircraft that can reach a speed of Mach 2 (1,320 mph [2,124km/h]).

In 1983, the first stealth fighters were introduced. The F-117A ground-attack fighter (below) was the first operational stealth aircraft.

The supersonic F-22 Raptor (below) has stealth capabilities. Early in the next millennium, it will replace the F-15 as America's dominant fighter.

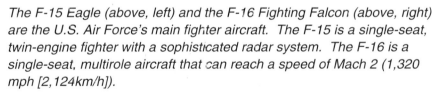

Jet Bombers

Germany built the first operational jet bomber in 1945. But it came too late to affect the outcome of World War II.

In the 1950s, the next generation of jet bombers, capable of delivering nuclear bombs great distances, first appeared in America. By 1960, the United States had developed the first supersonic bomber. But it was vulnerable to advanced surface-to-air missiles. A supersonic, low-altitude bomber that could avoid air defenses appeared in the late 1960s. But the airplane had a small payload.

The Soviet Union was the first to develop larger, long-range, low-level strategic bombers with more advanced avionics in 1975. America added its own version in the mid-1980s. In 1989, the United States introduced the first stealth bomber.

The German Arado Ar 234 (below) was the first operational jet bomber.

The Convair B-58 (above) was the first supersonic bomber.

America's B-52 Stratofortress (left) entered service in 1955. For the next 30 years, it remained the premier high-altitude, long-range bomber.

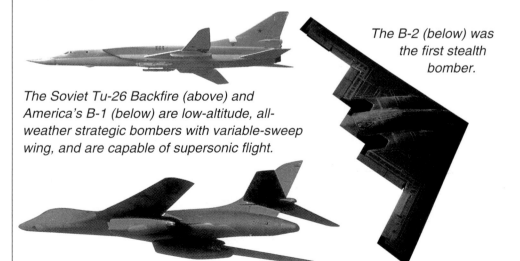

The B-2 (below) was the first stealth bomber.

The Soviet Tu-26 Backfire (above) and America's B-1 (below) are low-altitude, all-weather strategic bombers with variable-sweep wing, and are capable of supersonic flight.

Attack Helicopters

The first attack helicopters appeared in 1962 during the Vietnam War. They were small transport helicopters fitted with weaponry. The first helicopters built specifically as gunships entered service in 1967.

Gunships became tank killers in the 1980s as their weaponry became more lethal. Toward the end of the millennium, advanced avionics and fire-control systems allowed the attack helicopter to fight day or night and in all kinds of weather.

The UH1-Huey helicopters (left) were the first helicopters used as gunships. These transport helicopters were fitted with grenade launchers, rocket pads, and machine guns.

The AH-1G Huey Cobra (above) was the first helicopter designed as a gunship. Its stepped-up cockpit (pilot seat behind and above the gunner) revolutionized gunship design.

The Soviet Mi-24 Hind helicopter (right) was the first gunship with antitank missiles mounted on stub wings.

The AH-64 Apache (right) is the U.S. Army's main attack helicopter. It excels at killing ground targets, especially tanks. It can operate in bad weather and at night.

The Russian Ka-52 Alligator is a two-seat gunship, equipped with a wide array of weapons and capable of nighttime and bad weather missions.

TANKS

Tanks were invented in 1915 in England during World War I to break the stalemate caused by trench warfare.

The first tanks were adapted from tractors. They were slow, lightly armed, and had armor plating and caterpillar treads that helped them cross heavily defended German trenches along the Western Front. The newfangled war machines were shipped to France in crates marked "tank" to deceive German spies. The name stuck.

Tanks first saw battle on September 15, 1916, during the Somme offensive, and had modest success. But at the Battle of Cambrai in 1917, over 400 British tanks broke through enemy lines and helped capture 8,000 prisoners. The Allies used tanks successfully throughout the remainder of the war.

The British Mark I (above) was the first tank used in battle. It had light armor, machine guns, and an antibomb roof and tail, but it was heavy and slow with limited range.

The French Renault F.T. (left) was the first tank with a turret. Because of its speed and range, the Renault F.T. was the most widely used tank in World War I.

The M4 Sherman (left) was the principal tank of the U.S. Army during World War II. But its small 75-millimeter gun and light armor made it vulnerable to more powerful German tanks.

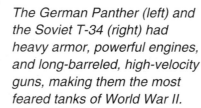

The German Panther (left) and the Soviet T-34 (right) had heavy armor, powerful engines, and long-barreled, high-velocity guns, making them the most feared tanks of World War II.

During World War II, the Germans were the first to organize tanks into fast-moving and powerful strike forces, a battle strategy that remains to this day. By then, tanks had become larger and more powerful, with heavy armor and large guns.

After the war, tank construction and weaponry continued to evolve as tanks became even larger. Powerful diesel and gas-turbine engines greatly increased speeds. Reactive and composite armor supplemented steel construction. In 1963, high-velocity, 120-millimeter guns were introduced. Computerized gun control, laser range-finders, infrared-imaging devices, and turret stabilizers dramatically increased their firing accuracy.

The M1A1 Abrams (right) is the main battle tank of the U.S. Army and Marine Corps. Despite weighing 126,000 pounds (57,154 kg), its gas-turbine engine makes it fast and mobile. The Abrams's armor plating is almost impenetrable. Its 120-millimeter smoothbore gun can kill most enemy tanks at very long range.

In 1975, the Russians introduced the 125-millimeter gun on their T-64 battle tank. The Russian T-72 main battle tank (left) has a 125-millimeter gun that is automatically loaded from a carousel-type magazine below the turret.

In 1232 at the siege of K'ai-feng, *Chinese defenders used "arrows of flying fire" (above) to ignite Mongol tents. These first solid-propellant rockets had tubular cases made of tightly wrapped paper coated with shellac and filled with black powder. They were attached to guide sticks.*

ROCKETS & MISSILES

Rockets were first used as weapons of war in China in the early thirteenth century. But they did not become a deadly force until World War II when they were used as tank killers and artillery weapons. Missiles, rockets with guidance mechanisms, also appeared.

After the war, rockets and missiles became larger, more powerful, and sophisticated. And they could carry a variety of warheads.

Toward the end of the millennium, the largest missiles became the world's deadliest weapons of war. They could strike targets thousands of miles away with multiple nuclear warheads.

Early Rockets and Missiles

After the Chinese first used rockets in the thirteenth century, they quickly appeared in Europe. Steady improvements were made to rocket construction. But rockets were used only occasionally, and had limited effect on wars.

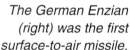

The German Enzian (right) was the first surface-to-air missile.

Rockets (right) were used by the British against the American Fort McHenry during the War of 1812, and are referred to in "The Star Spangled Banner."

A landing craft fires barrage rockets (right) during World War II.

Metal casings were added in the late eighteenth century in India, increasing the thrusting power and range. Sir William Congreve of England standardized rocket construction and added an explosive warhead. In the mid-nineteenth century, William Hale eliminated the guide stick by adding jet vents that spun the rocket and kept it stable during flight. The rocket became an effective weapon. But its use remained sporadic.

During World War I, rockets were used in trench warfare and they were used to shoot down dirigibles. But they were used little elsewhere.

During World War II, the rocket became a major weapon of war. Barrage rockets could saturate enemy ground positions with devastating effect. And shoulder-mounted rockets proved effective tank killers. The forerunners of the cruise missile and the ballistic missile also appeared.

The V-1 (below) was a fixed-wing missile propelled by a pulse jet. The V-1 carried 2,000 pounds (907 kg) of explosives, but was slow and defensible.

The German Nebelwerfer (right) was a six-tube barrage rocket launcher that U.S. and British troops called the "Screaming Meemie" for the eerie sound the rockets made.

During World War II, the Allies developed the antitank rocket, called the bazooka (left). It was fired by a soldier from a shoulder-held tube. It could hit tanks up to 200 yards (182 m) away.

The V-2 (right) was the first ballistic missile. It had an internal guidance and control system to deal with in-flight rolling, pitching, and yawing. V-2s carried 1,650 pounds (748 kg) of explosives and traveled faster than the speed of sound, making them indefensible.

German scientist Wernher von Braun (left) was the father of the V-2 missile. After the war, he helped the U.S. develop its strategic missile program.

The Soviet SA-7 Grail was the first shoulder-fired, heat-seeking missile.

In 1953, Nike Ajax (above) was the first guided surface-to-air (SAM) system in the world to enter operational service.

The Snark (right) was the first U.S. cruise missile. It traveled at 600 mph (965 km) with a range of 5,000 miles (8,045 km), and carried a 2,000-pound (907 kg) atomic warhead.

Modern Missiles

After the war, a host of guided missiles were developed because of advancements in electronics. Wire-guided and radar-guided missiles appeared in the late 1940s. Heat-seeking missiles were developed in the 1950s. So, too, were the first SLBMs. In 1959, the first ICBMs were deployed.

Antiradar missiles were used extensively by the U.S. in the Vietnam War. Late in the 1960s, the first MRVs were added to long-range ballistic missiles.

Missile technology continued to develop, as missile ranges increased. MIRVs first appeared in 1970. And laser-guided and optically-guided missiles were developed in the early 1970s. In the 1980s, the first successful strategic cruise missiles entered service. By the end of the millennium, missiles had evolved into air-to-air, air-to-ground, surface-to-air, and surface-to-surface weapons. They were tank, ship, submarine, and missile killers.

Deployed in 1970, the U.S. TOW (above) is a wire-guided missile. As the gunner keeps the cross hairs on the target, a computer in the launcher sends flight instructions to the missile through the wires.

Developed in 1963, the Shrike (above) was the first antiradiation air-to-ground missile used in combat. It homed in on enemy radar.

The U.S. Tomahawk long-range cruise missile (above) flies from 50 to 100 feet (15 - 30 m) overhead as the missile's radar compares the terrain underneath to a three-dimensional map in the onboard computer. Cruise missiles are very precise. They can hit single objects, like tanks or bunkers, hundreds of miles away.

The AIM-9 Sidewinder (right) was the first heat-seeking missile.

The Patriot (above) was the first successful antimissile system.

The AGM-114 Hellfire missile (left) is the main armament of several U.S. attack helicopters. They are laser-guided missiles used mainly to kill tanks. They travel faster than the speed of sound, with a range of about 5 miles (8 km).

The U.S. Harpoon (left) is a sea-skimming antiship missile. It comes in low to avoid radar detection, then performs a sudden pop-up and dive onto the target ship to defeat automatic rapid-fire defense systems. Its range is approximately 58 miles (93 km), powered by a small turbojet engine.

The U.S. MX Peacekeeper (below) is a three-stage ICBM that carries 10 300-kiloton warheads and has a range of 7,000 miles (11,263 km).

The Polaris (below), first launched in 1960, was America's first SLBM. The Soviets launched the first-ever SLBM in 1955.

A Multiple Independent Re-entry Vehicle (MIRV—above) has small, black, cone-shaped warheads that can deliver to separate targets.

Deployed in 1959, the Atlas (above) was one of the first ICBMs.

WEAPONS OF MASS DESTRUCTION

Biological weapons were the first weapons used in war that held the potential for mass destruction. But the lack of scientific knowledge prevented successful use. Toward the end of the millennium, many countries agreed to stop developing biological weapons.

Chemical weapons were used successfully on a large scale during World War I. But effective defenses ultimately limited their overall impact on the war. Since that time, chemical weapons have been used sparingly in only a few wars.

The first effective weapon of mass destruction was the nuclear bomb which brought an end to World War II. Nuclear bombs led the way for development of thermonuclear warheads that sit atop ballistic missiles found in modern nuclear arsenals around the world.

Biological Warfare

At the siege of Caffa in 1347, the Mongols infected their enemies, the Genoese, with the bubonic plague by catapulting infected bodies over the city walls. But a basic understanding of microbes was not known, which limited the success of this early biological warfare tactic.

Modern biological weapons have never been used on a large scale in warfare. They consist of infectious microbes, including bacteria, viruses, and fungi, for use against people, animals, and plants. Released into the air, microbes could kill large, unprotected populations. In 1972, a treaty was signed by the U.S. and seventy other countries that banned the development, production, and stockpiling of biological weapons.

Chemical Warfare

After spending years in desperate trench fighting, the Germans tried to break the stalemate by using chlorine gas and phosgene, which hurt the lungs, and mustard gas, which causes severe burns. But protective devices such as gas masks and overgarments made the chemicals ineffective.

A U.S. Marine wears a protective biohazard suit during a training exercise simulating a biological weapons attack.

Chemical weapons were not used during World War II, even though opposing sides stockpiled them. Their most recent use, during the Iran-Iraq war in the 1980s, also had little effect on the war's outcome.

Modern chemical weapons such as sarin, VX, and soman use toxic nerve agents, which cause breathing difficulties and skin damage. These deadly liquids are delivered to targets in artillery shells, missiles, rockets, bombs, and other munitions.

Germany initiated the first poison gas attack in April 1915, during World War I. The deadly chlorine was released from thousands of cylinders.

The first nuclear bomb test, known as Trinity, occurred on July 16, 1945, at the Alamogordo Bombing Range in New Mexico.

Nuclear Warfare

In 1945, the United States used an atomic bomb against Japan to bring a quick end to World War II. The first atomic bombs were fission bombs, which split atoms to unleash the power of matter itself.

After the war, fusion bombs were developed. Fusion bombs use the same kind of thermonuclear fuel that powers the sun. Thermonuclear warheads were soon placed atop ballistic missiles. Over the years, warheads became more powerful and diverse. At the end of the millennium, nuclear arsenals also included artillery shells, torpedoes, depth charges, land mines, cruise missiles, and short-range missiles.

On August 6, 1945, "Little Boy" (left)—the first atomic bomb used in war—was detonated over the city of Hiroshima, Japan (right). The bomb had the power of about 20,000 tons (20,320 t) of TNT, and killed 90,000 people.

On November 1, 1952, a 10.4-megaton thermonuclear explosion, code-named MIKE (right), ushered in the thermonuclear age. The test site—the island of Elugelab—was completely vaporized.

WEAPONS

CLOSE COMBAT

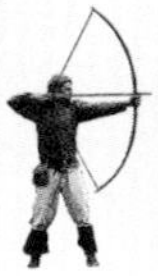

First longbows appear in Europe

First metal crossbows

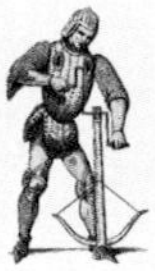

Sword hand guards developed (1350s); pike, halberds appear

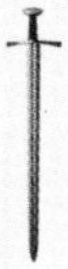

SHIPS

First galleons developed

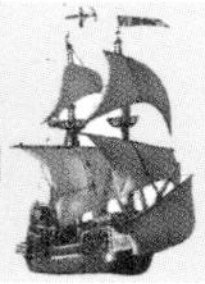

GUNS & ARTILLERY

First cannon developed

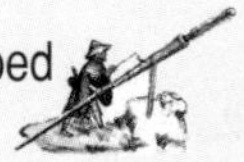

Wrought-iron barrels appear on cannon; first firearms developed

| 1000 | 1100 | 1200 | 1300 | 1400 | 1500 |

AIRPOWER

First fighter aircraft developed (1915)

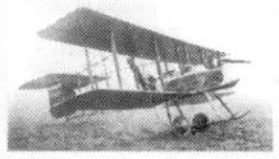

First operational jet-powered aircraft (1939)

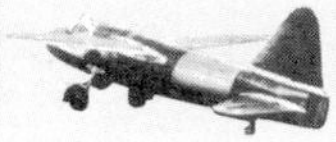

First operational jet bomber (1945)

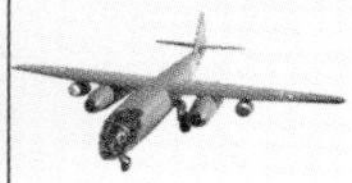

TANKS

First tank (1915); first tanks used in battle (1916)

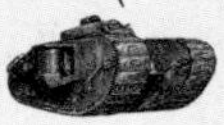

ROCKETS & MISSILES

China uses first rockets (1200s); India adds metal cases to rockets (1700s)

First ballistic missile developed (1936)

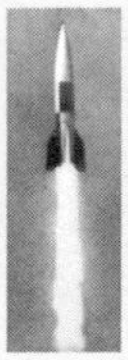

First wire- and radar-guided missiles

WEAPONS OF MASS DESTRUCTION

Mongols use biological warfare (1347)

First poison gas attack (1915)

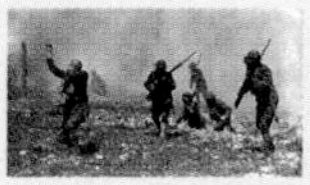

First nuclear bomb test and use in war (1945)

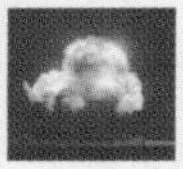

| Pre-1900 | 1900 | 1910 | 1920 | 1930 | 1940 | 1950 |

MILESTONES

Bayonet invented (1640s); socket bayonet invented (1688)

SHIPS

First man-of-war; first submarine used in war (1770s)

First battleships, ironclad ships, destroyers, and cruisers launched

First aircraft carriers developed; first nuclear-powered ships appear

GUNS & ARTILLERY

Firearms replace bows and arrows; cannon made of cast iron and bronze

First machine gun patent (1718)

First revolvers, breechloading cannon, successful machine guns

First submachine guns, automatic rifles appear

1500	1600	1700	1800	1900	2000

AIRPOWER

First supersonic fighters, strategic bombers appear

Mach 2 speed broken; first supersonic bombers; first helicopter gunships (1967)

First stealth fighters (1983) and stealth bombers (1989)

TANKS

First gas-turbine engines; 120 mm guns appear

125 mm guns appear; first automatic loading mechanisms

ROCKETS & MISSILES

First heat-seeking missile, ICBM, and SLBM

First MRV and antiradiation missile

First shoulder fired missiles; laser- and optically-guided missiles and MIRVs developed

First strategic cruise missile; first antimissile system

WEAPONS OF MASS DESTRUCTION

First hydrogen bomb test (1952)

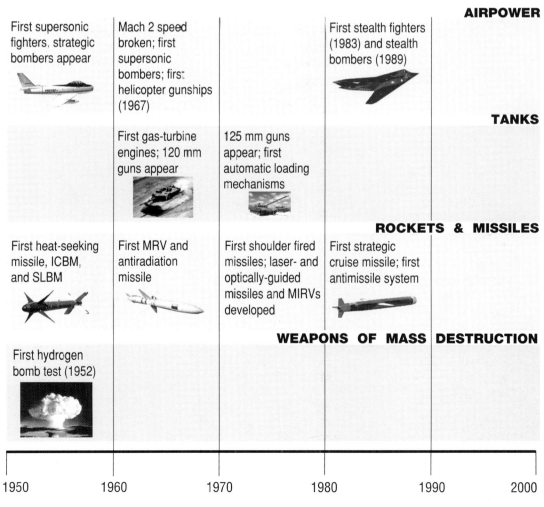

1950	1960	1970	1980	1990	2000

GLOSSARY

aerodynamic - a design that reduces air resistance. It allows automobiles, trains, and aircraft to travel faster and use fuel more efficiently.

after-burning engine - an engine that gets additional thrust when afterburners inject fuel into its hot exhaust gasses.

ammunition - bullets, shells, and other projectiles used in firearms and artillery.

amphibious - coordinating land, sea, and air forces for invasion.

antiradiation - a missile that seeks out and destroys radar systems.

arsenal - a collection of weapons.

artillery - weapons for discharging missiles.

assault rifle - automatic or semiautomatic rifles designed for military use that have large magazines.

automatic - a firearm that loads a round, fires it, and ejects the spent cartridge in one cycle that begins when the trigger is pulled and continues as long as the trigger is held down. Semiautomatic firearms also load, fire, and eject in one cycle, but each cycle must be started by an individual pull of the trigger.

avionics - electronics designed for aerospace vehicles.

B.C. - a designation of time, used in expressing dates before the birth of Christ. The abbreviation B.C. stands for "before Christ."

bacteria - tiny living microbes that can only be seen through a microscope. Some bacteria are bad and cause diseases, and some are good, such as the ones that turn milk into cheese.

barrage - a vigorous or rapid projection of many missiles at once.

barrel - the tube on a gun that the bullet comes out of.

battering ram - a military device used in ancient times to beat down the doors and walls of a besieged place.

biohazard - a word that combines biological and hazardous. A substance that can cause harm to people, animals, or the environment.

biological - something that relates to living and life processes.

blitzkrieg - a war conducted with great speed and force consisting of a violent surprise attack by coordinated air and ground forces.

bolt action - a firearm that is loaded by a bolt, which is a cylindrical shaped metal piece that introduces the cartridge into the chamber.

bombard - to assault vigorously or persistently.

bore - the inside of a gun barrel. Smooth bores have no marks. Rifled bores are etched with a spiral pattern that causes the bullet to spin, which makes it travel further and straighter.

breechblock - a mechanism on a firearm that closes the breech against the force of the explosion and prevents gas from escaping.

Bronze Age - a period in human history when bronze was used to make tools and weapons.

bubonic plague - a serious disease caused by bacteria that is passed from one person to another through the air. People with bubonic plague have swollen lymph glands and a high fever.

cartridge - a cylindrical case, usually made of metal, that combines a percussion cap, gunpowder, and a bullet in one container.

catapult - an ancient military device for hurling projectiles.

caterpillar tread - two endless metal belts that can propel a vehicle over rough or soft terrain.

cavalry - an army division that fights on horseback.

chamber - a compartment in the cylinder of a gun that holds a cartridge.

cock - to pull the hammer of a firearm back to prepare the trigger for firing.

combustion - the act of burning. The gas released by combusting gunpowder is harnessed to power the load-fire-eject cycle of some automatic and semiautomatic weapons.

commissioned - the act of introducing a ship into service. When a ship is taken out of service, it is decommissioned.

composite - something that is made up of different, distinct parts.

convoy - a group that travels together.

coolant agent - a fluid that keeps an engine or machine from overheating.

crossbow - a weapon that consists of a small bow mounted crosswise on the end of a stock.

cruise missile - a guided missile that flies at moderate speed and low altitude and has a terrain-following radar system.

cylinder - the part on a firearm that holds the cartridges.

deploy - to place in battle position.

depth charge - an antisubmarine weapon that consists of a barrel filled with explosives that is dropped into the water where it explodes at a predetermined depth.

diesel engine - an internal combustion engine that uses hot, pressurized air to ignite the fuel.

dogfight - a fight between two or more aircraft at close range.

ejector rod - a mechanism on a firearm that ejects or pushes out a used cartridge.

exotic - different or unusual.

firearm - a weapon, usually small, through which a bullet is discharged by gunpowder.

flammable - something that can be easily ignited and burns fast.

forecastle - the forward part of the upper deck of a ship.

fortify - to add material to something to strengthen it.

galleon - a heavy, square-rigged sailing ship used for war or commerce.

hammer - the part of a firearm that strikes the cartridge or percussion cap and ignites the gunpowder.

heat-seeking - a missile that aims for a target based on the target's temperature.

hilt - the handle of a sword or dagger.

ICBM - Intercontinental Ballistic Missile. ICBMs are missiles that are capable of traveling between continents. They follow a high upward trajectory and free fall in descent.

ignition - the mechanism used to start gunpowder or fire.

infantry - soldiers that are trained, armed, and equipped to fight on foot.

infrared - radiation with wavelengths longer than those of visible light, but shorter than radio waves. This radiation can be measured to determine the temperature of an object. Infrared imaging creates a thermal picture of a target.

Iron Age - the period in human history after the Bronze Age, in which iron was smelted and used to make tools and weapons.

knapped stone - stone that has been chipped and shaped.

laser-guided - a missile that is directed toward its target by a laser.

laser range finder - laser technology that measures the range of a target.

lethal - capable of causing death.

longbow - a hand-held wooden bow that is held vertically.

magazine - a holder in or on a firearm that contains the cartridges that will be fed into the chamber.

maneuverability - to easily execute tactical movements.

medieval - something that is of or has characteristics typical of the Middle Ages. The Middle Ages date from around A.D. 500 to 1500.

microbe - a tiny organism that can only be seen through a microscope. A germ.

mine - an explosive charge placed underground or underwater that explodes when disturbed. A minelayer is a ship that plants mines underwater. A minesweeper is a ship that finds, removes, or destroys underwater mines.

MIRV - Multiple Independent Re-entry Vehicle. Several nuclear warheads carried on the front of a ballistic missile. An MIRV can independently launch several warheads at different targets.

misfire - failure to fire or ignite at the proper time.

missile - a rocket-propelled weapon that launches warheads with great accuracy and speed. A guided missile has a flight path controlled by electric signals or other means. Cruise missiles are a type of guided missile that flies at moderate speed and low altitude, with a terrain-following radar system. Ballistic missiles are self-propelled, with a controlled ascent and a free-fall descent.

MRV - Multiple Re-entry Vehicle. Several warheads from the same missile that can be launched at one target, increasing the chances of striking it.

munitions - military supplies, such as guns, ammunition, or bombs.

nitroglycerin - a heavy, oily, poisonous liquid used in making explosives.

nose-intake - the opening in the nose of a jet that allows air to enter the engine. The air mixes with fuel to create thrust, which powers the aircraft.

optically-guided - something that is guided by sight.

payload - the load carried by a vehicle that is not necessary for the vehicle's operation.

percussion cap - a paper or metal container holding an explosive charge.

pitch - a black, sticky substance made of petroleum or tar that is used for waterproofing.

poop deck - the short deck above the ship's main stern deck, often forming the roof of a cabin.

projectile - an object that is designed to be projected through space, such as a bullet.

prone - likely to happen.

radar-ranging gunsights - radar technology that allows a gunner to lock in on a target.

recoil - to spring back from the force of impact or discharge.

reconnaissance - a survey to gain information, especially military information about enemy territory.

retractable landing gear - landing gear is the part of the airplane that supports its weight when it is on the ground. Retractable landing gear moves out of the plane's fuselage when it lands and goes back in after the plane takes off.

sheath - a case for the blade of a sword, knife, or dagger.

silo - underground containers where some kinds of missiles, such as ICBMs, are launched.

SLBM - Submarine-Launched Ballistic Missile. An underwater missile launched by a submarine.

stalemate - a situation when no further action is possible; a standstill.

steam-turbine engine - a rotary engine that gets power from steam passing by vanes on a spindle.

stockpile - to put aside supplies for an emergency or shortage.

stub wing - the devices on the side of a helicopter that carry weapons.

supersonic - transportation that moves from one to five times the speed of sound.

sweepback - the backward slant of an airplane wing that makes the tip of the wind downstream from the part that is attached to the fuselage. This reduces air resistance and increases speed.

technology - the use of scientific knowledge to solve practical problems, especially in industry.

terrain - a region of land, with regard to its natural features.

thermonuclear - the fusion of atomic nuclei at temperatures of millions of degrees, as in a hydrogen bomb.

through-deck aircraft - the length of an aircraft carrier that is devoted to the landing and taking off of aircraft.

toggle - a device used for holding or securing.

torpedo - a propeller-driven, cigar-shaped missile used underwater to rupture a ship's hull.

turbojet - a jet engine in which a turbine drives a compressor that supplies air to a burner. Hot gases from the burner drive the turbine before being discharged rearward.

vaporize - to change something into a vapor.

velocity - speed.

versatile - having many uses or functions.

virus - a small organism that can reproduce and grow only in living tissue. Viruses can cause diseases in humans, animals, and plants.

vulnerable - easily damaged or hurt.

warfare - the act of fighting a war.

warhead - the front part of a torpedo or missile that carries the explosive charge.

winch - a mechanism for pulling or lifting.

windlass - a kind of winch turned by a hand crank.

wrought iron - a pure form of iron that is tough, easily worked and welded, and relatively resistant to corrosion.

46

INTERNET SITES

Armed Forces of the World
http://www.cfcsc.dnd.ca/links/milorg/index.html
This site contains links to military related web sites from nearly every country in the world. A good starting point for collecting raw information and hard-to-find statistics.

The Aviation History On-line Museum
http://www.aviation-history.com/
This informative site includes details of important aircraft and pilots, including the early years of aviation history.

United States Naval & Shipbuilding Museum
http://www.uss-salem.org/
This online museum includes information on naval history and shipbuilding techniques, with over 6,000 ship histories included.

Nuclear War
http://tqd.advanced.org/3471/noNetscape/nuclear_war.html
This informative site divides the topic of nuclear war into three sections: weapons, warfare, and politics. Read about missiles, carrier devices, terrorists, and the aftermath of nuclear war. It presents opposing viewpoints of the major issues of nuclear war.

These sites are subject to change. Go to your favorite search engine and type "weapons" for more sites.

FOR FURTHER READING

Black, Wallace B. and Jean F. Blashfield. *Blockade Runners and Ironclads: Naval Action in the Civil War.* Danbury, Connecticut: Franklin Watts, Inc., 1997.

Byam, Michele. *Arms and Armor.* New York: Alfred A. Knopf, Inc., 1988.

Humble, Richard. *A World War Two Submarine.* Lincolnwood, Illinois: NTC Publishing Group, 1991.

Meltzer, Milton. *Weapons & Warfare: From the Stone Age to the Space Age.* New York: Harper Collins, 1996.

Otfinoski, Steven. *Blasting Off: Rockets Then and Now.* London: Marshall Cavendish Corp., 1998.

Young, Robert. *Hiroshima: Fifty Years of Debate.* Severna Park, Maryland: Silver Burdett Press, 1994.

INDEX